My First Science Words

MATTER WORDS

A CRABTREE SEEDLINGS BOOK

Taylor Farley

A Crabtree Seedlings Book

Crabtree Publishing
crabtreebooks.com

matter

(MAT-ur)

solid

(SOL-id)

liquid

(LIK-wid)

gas

(GAS)

states
(STAYTS)

gas
solid
liquid

describe

(dehs-KRYB)

long,
heavy,
scaly . . .

color

(KUHL-ur)

green
blue
yellow
orange
pink

size

(SYZE)

big

small

texture
(TEKS-chur)

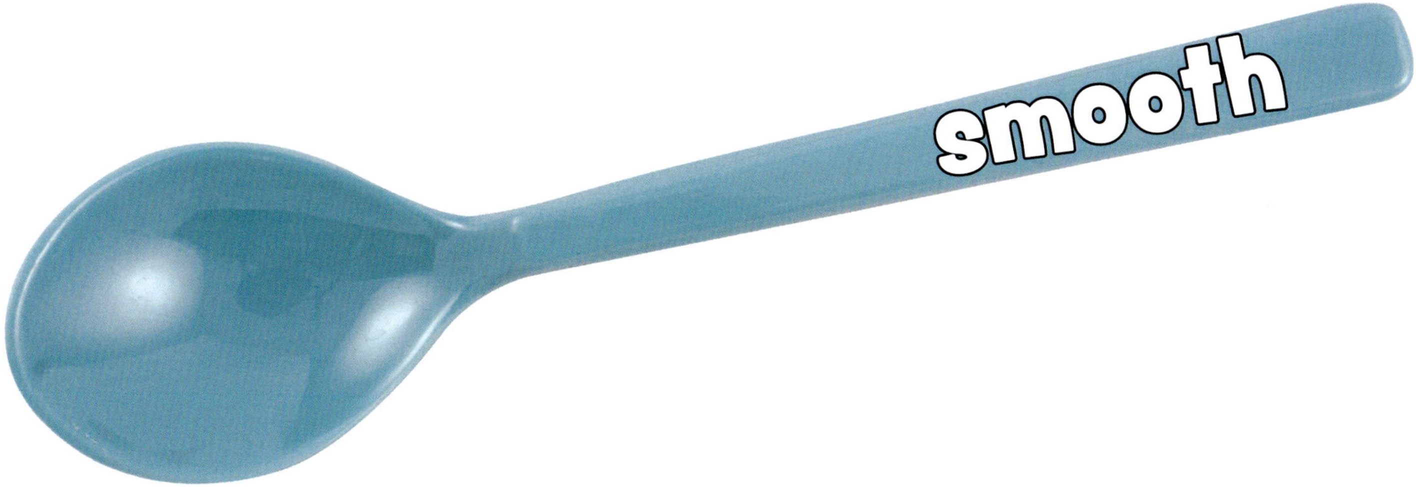
smooth

bumpy

shape
(SHAYP)

weight
(WAYT)

Glossary

color (KUHL-ur): Color is what we see when light shines on objects. We give color names, such as red, white, and yellow.

describe (dehs-KRYB): When you describe something, you tell about it in words.

gas (GAS): Gas is matter that we usually cannot see. Air is a gas.

liquid (LIK-wid): Liquid is matter that is wet and can be poured. Juice is a liquid.

matter (MAT-ur): Matter is anything that has weight and takes up space. Matter is all around us.

shape (SHAYP): Shape is the form or outline of an object or figure. A ball has a round shape.

size (SYZE): Size describes how big something is. We can measure size in different ways.

solid (SOL-id): A solid is matter that has a shape and is not a liquid or a gas.

states (STAYTS): States are different forms, or types, of matter. Ice is the solid state of water.

texture (TEKS-chur): Texture is the look and feel of something. A plastic spoon has a smooth texture.

weight (WAYT): Weight is a measure of how heavy something is.

School-to-Home Support for Caregivers and Teachers

Crabtree Seedlings books help children grow by letting them practice reading. Here are a few guiding questions to help the reader build his or her comprehension skills. Possible answers are included.

Before Reading

- **What do I think this book is about?** I think this book is about types of matter.
- **What do I want to learn about this topic?** I want to learn about how types of matter are similar and different.

During Reading

- **I wonder why...** I wonder what other words the boy on page 13 could use to describe the snake.
- **What have I learned so far?** I have learned that words related to color, size, texture, shape, and weight tell us about how matter is similar and different.

After Reading

- **What details did I learn about this topic?** I learned that the three states of matter are solid, liquid, and gas.
- **Write down unfamiliar words and ask questions to help understand their meaning.** I see the word ***describe*** on page 12 and the word ***texture*** on page 18. The other vocabulary words are found on pages 22 and 23.

Crabtree Publishing

crabtreebooks.com 800-387-7650

Hardcover 978-1-4271-3044-0
Paperback 978-1-4271-3049-5

Printed in Canada/122023/CPC20231207

Published in Canada
Crabtree Publishing
616 Welland Avenue
St. Catharines, Ontario
L2M 5V6

Published in the United States
Crabtree Publishing
347 Fifth Avenue
Suite 1402-145
New York, NY 10016

Written by: Taylor Farley

Photo Credits: Cover and page 5 legos © ESOlex, cover soda bottles © AlenKadr, cover and page 11 ice cube and gas illustrations © GraphicsRF, cover glass of water illustration © Lina Truman, cover and page 9 balloons illustration © Yganko, photo page 3 © FamVeld, page 7 juice © Theeradech Sanin, page 13 snake © shutterstock.com/ BLUR LIFE 1975 Editorial credit: Ekaterina Malskaya, page 15 colored powder © jkjainu, page 17 © dogs Erik Lam, page 19 spoon © onair, bumpy texture © Korelidou Mila, page 21 © wacomka, page 23 © Jesse Davis. All photos from Shutterstock.com. Page 24 © shutterstock.com/r.classen

Library and Archives Canada
Cataloguing in Publication
Available at the Library and Archives Canada

Library of Congress
Cataloging-in-Publication Data
Available at the Library of Congress